Petar Piljek
Zdenka Keran
Ante Ninić

Micromachining

Petar Piljek
Zdenka Keran
Ante Ninić

Micromachining

State of the Art

LAP LAMBERT Academic Publishing

Impressum / Imprint

Bibliografische Information der Deutschen Nationalbibliothek: Die Deutsche Nationalbibliothek verzeichnet diese Publikation in der Deutschen Nationalbibliografie; detaillierte bibliografische Daten sind im Internet über http://dnb.d-nb.de abrufbar.
Alle in diesem Buch genannten Marken und Produktnamen unterliegen warenzeichen-, marken- oder patentrechtlichem Schutz bzw. sind Warenzeichen oder eingetragene Warenzeichen der jeweiligen Inhaber. Die Wiedergabe von Marken, Produktnamen, Gebrauchsnamen, Handelsnamen, Warenbezeichnungen u.s.w. in diesem Werk berechtigt auch ohne besondere Kennzeichnung nicht zu der Annahme, dass solche Namen im Sinne der Warenzeichen- und Markenschutzgesetzgebung als frei zu betrachten wären und daher von jedermann benutzt werden dürften.

Bibliographic information published by the Deutsche Nationalbibliothek: The Deutsche Nationalbibliothek lists this publication in the Deutsche Nationalbibliografie; detailed bibliographic data are available in the Internet at http://dnb.d-nb.de.
Any brand names and product names mentioned in this book are subject to trademark, brand or patent protection and are trademarks or registered trademarks of their respective holders. The use of brand names, product names, common names, trade names, product descriptions etc. even without a particular marking in this works is in no way to be construed to mean that such names may be regarded as unrestricted in respect of trademark and brand protection legislation and could thus be used by anyone.

Coverbild / Cover image: www.ingimage.com

Verlag / Publisher:
LAP LAMBERT Academic Publishing
ist ein Imprint der / is a trademark of
OmniScriptum GmbH & Co. KG
Heinrich-Böcking-Str. 6-8, 66121 Saarbrücken, Deutschland / Germany
Email: info@lap-publishing.com

Herstellung: siehe letzte Seite /
Printed at: see last page
ISBN: 978-3-659-61510-8

Copyright © 2014 OmniScriptum GmbH & Co. KG
Alle Rechte vorbehalten. / All rights reserved. Saarbrücken 2014

Contents

1. Introduction 3
2. Micromachining process physic 7
   2.1. Size effects 7
   2.2. Workpiece material 8
   2.3. Minimum chip thickness 10
   2.4. Cutting forces 12
   2.5. Brittle and ductile mode machining 16
   2.6. Surface quality 16
3. Micro Cutting Tools 20
   3.1. Diamond tools 21
   3.2. Tungsten carbide (WC) tools 22
   3.3. Coatings 23
   3.4. Tool manufacturing methods 24
   3.5. Tool failure 25
   3.6. Tool design 26
4. Machine tools with micromachining capability 27
   4.1. Ultra-precision machine tools and micromachine centres 27
      4.1.1. Machine materials 28
      4.1.2. Spindle bearings and linear guides 28
      4.1.3. High resolution of linear and rotary motions 29
      4.1.4. Computer Numerical Control (CNC) 30
      4.1.5. Position measurement and process monitoring 31
   4.2. Miniaturized machine tools and micro factories 32

5. Numerical modelling of micromachining 34
   5.1. FEM 34
   5.2. MD 38
   5.3. FEM and MD 41
6. Conclusion.. 42
7. Bibliography... 45

1. Introduction

The trend of micro-miniaturization of the products and its parts has already become forceful in industry, especially in field of micro electromechanical system (MEMS) or micro system technology (MST). In MEMS manufacturing techniques such as *photolithograpy, chemical-etching, plating* and *LIGA* are used, as shown in Figure 1. They are very well known in semiconductors or microelectric manufacturing and used for large volume production, mainly sensors and actuators made of silicon or limited range of metals.

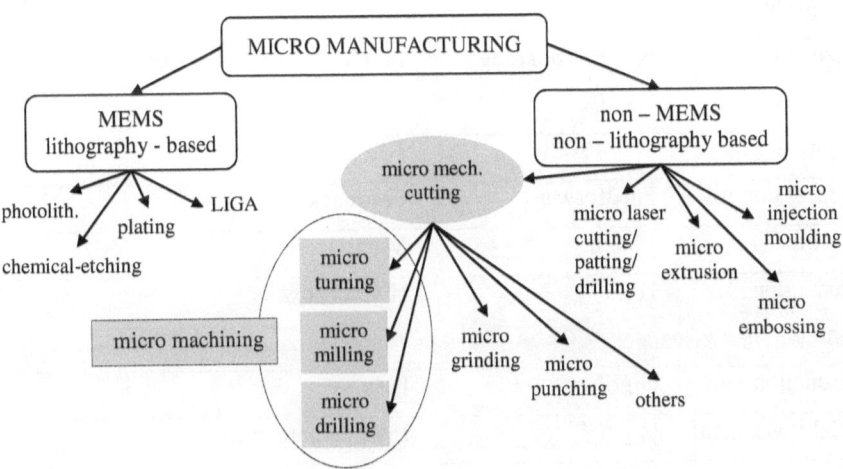

Figure 1: Classification of micro manufacturing techniques.

However, in the last two decade new category of micro manufacturing techniques have been developed, known as non-MEMS or non-lithography-based micro manufacturing. Non-lithography-based micro manufacturing include techniques such as *micro EDM, micro mechanical cutting, micro laser cutting/patting/drilling, micro extrusion, micro embossing, micro stamping* and *micro injection moulding*, (Figure 1) and these manufacturing techniques are fundamentally different from MEMS micro manufacturing in many aspects [1]. Non-lithography-based micro manufacturing can produce high-precision three dimensional products using a variety of materials and possessing features with size ranging from tens of micrometres to a few millimetres.

Table 1 shows the fundamental differences between MEMS micro manufacturing and micromachining.

Table 1: Comparisons between MEMS – based process and micro machining (adapted from [1]).

	MEMS – based process	Micro mechanical machining
Workpiece materials	Silicon, some metals	Metals, alloys, polymers, composite, technical ceramics
Component geometry	Planer or 2.5D	Complex 3D
Assembly methods	None or bonding	Fastening, welding, bonding
Relative accuracy	$10^{-1} - 10^{-3}$	$10^{-3} - 10^{-5}$
Process control	Feedforward	Feedback
Machine size	Macro	Macro or micro
Production volume	High	High or low
Production rate	High	Low
Total investment	High	Intermediate or low
Applications	MEMS, microelectronics, some planner micro parts	Various applications requiring 3D micro components

Micromachining refers to mechanical micro cutting using geometrically determined cutting edge(s) (micro turning, micro milling and micro drilling, etc.) performed on conventional precision machines or micromachines. Although lithography-based manufacturing can achieve smaller feature size, micromachining has many advantages in terms of material choices, relative accuracy and complexity of produced geometry. Moreover, it is a promising technology for bridging the gap between macro and nano/micro domain [1,2], as can be seen in Figure 2.

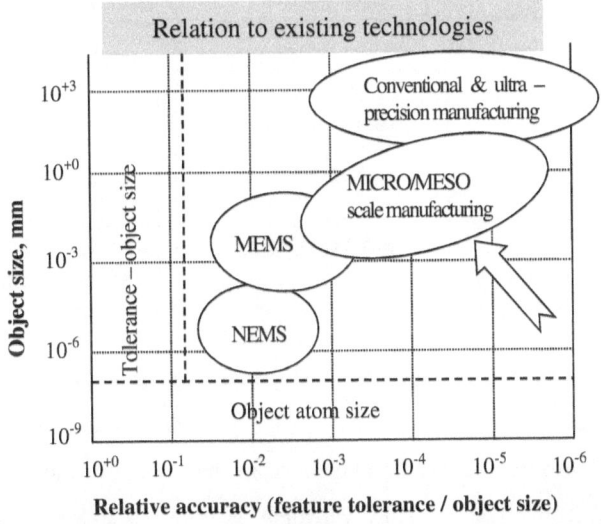

Figure 2: Micro manufacturing size/precision domains (adapted from [1]).

Although micromachining techniques are similar to conventional (macro) machining manufacturing techniques, simple scaling of parameters or process model cannot be applied due to size effects. There are two research approaches taken to deal with size effects. These two approaches overlap in some areas and attempt to address similar issues, such as cutting tool edge size effect, minimum chip thickness, and so on [1]. One approach is based on minimization of the conventional machining process, tooling and equipment with an emphasis on their scaling down effects. Macro models are adapted to micro cutting with consideration of the size effects. The other approach, covered in this paper, find its origin in ultra-precision machining, with the emphasis on cutting mechanics. This approach is similar to diamond cutting research, but studies micro cutting, with more emphasis on tool geometries, material crystalline orientation and micro structures. Key aspects that have influence on micromachining process are shown in Figure 3.

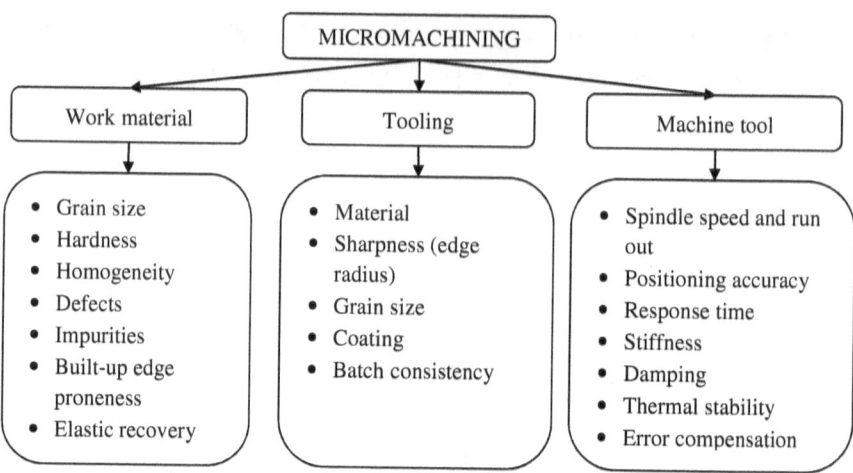

Figure 3: Key aspects in micromachining (adapted from [3]).

Although research in micro cutting has been reported since late sixties [3,4] strong interest in micromachining can be evident from the middle of the last decade, as it can be noticed from Figure 4. However, there is little research papers dealing with materials that cannot be machined easily [3]. Micromachining of materials such as hardened steels, stainless steels, silicones, glasses and ceramics introduces additional problems related to excess tool wear, unpredictable tool failure, low stiffness of the micro tools, surface and subsurface cracks, etc.

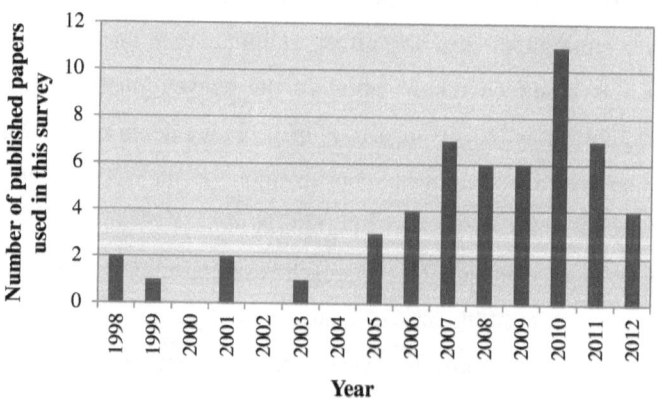

Figure 4: Evolution of the number of papers published on micromilling in recent years (adapted from [3]).

The paper is divided into four main parts which are dealing with process physic, micro cutting tools, micro machine tools and numerical modelling, and within them subjects such as size effects, workpiece material requirements, surface quality, cutting tool material, geometry, wear and failure mechanisms, machine tools, sensors and other related technicalities are discussed.

2. Micromachining process physic

2.1. Size effects

Size effects are certainly among the principal issues, if not the most relevant aspect, to be addressed in micromachining [5,6,3,2,7,8]. It is typically characterised by a dramatic and non-linear increase in the specific energy (energy consumed per unit volume of material removed) as the undeformed chip thickness decreases. Experimental observation of this phenomenon in machining of ductile metal (SAE 1112 steel) has been reported in early work by Backer et al. [9]. They performed a special series of tests to determine relation between shear stress and chip thickness. The results from the experiment where later modified by Taniguchi [10] as shown in Figure 5.

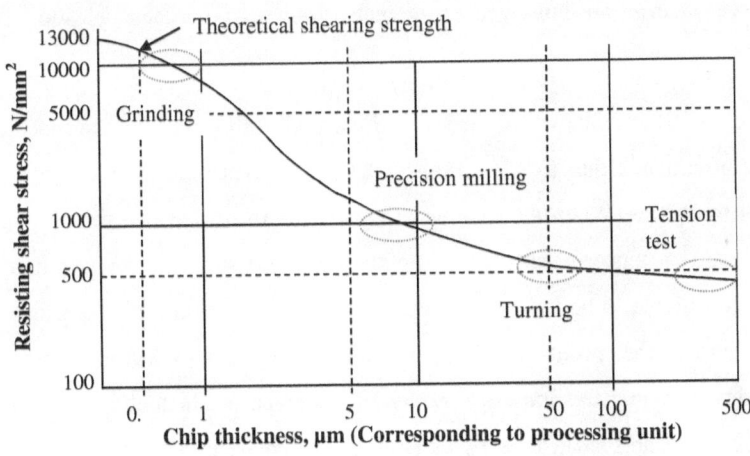

Figure 5: Relation between chip thickness and resisting shear stress modified by Taniguchi.

Although micromachining includes many characteristic of conventional (macro) machining process, the size effect modifies the mechanism of material removal and prevents the production parameters to be changed according to the rules of similarity. There are two different aspects of size effects of concern, when the thickness of material to be removed is of the same order of magnitude as the tool edge radius, or where the microstructure of workpiece material has significant influence on the cutting mechanism [6].

The size effect was attributed to tool edge radius effect, material micro-structure effect i.e. dislocation density/availability, crystallographic orientation, material strengthening effect due to strain, strain rate, strain gradient, subsurface plastic deformation, material separation effect and cutting speed. However, there is no clear agreement on the origin of the size effect [8].

2.2. Workpiece material

In conventional machining workpiece is often considered to be homogeneous and isotropic. Such an assumption cannot be made when dealing with micromachining processes due to size effects caused by workpiece material microstructure. As evident from the Figure 3, key aspects to be considered in micromachining related to workpiece material are homogeneity, defects, grain size, hardness, elastic recovery, etc.

Backer et al. [9] and Shawn [5] discusses the origin of the size effect in metal cutting which consequence due to short range inhomogeneities present in all commercial engineering metals. When the volume of material deformed at one time is relatively large, there is a uniform density of imperfections and strain (and strain hardening) may be considered to be uniform. However, as the volume deformed approaches the small volume, the probability of encountering a stress-reducing defect (grain boundaries, missing and impurity atoms, etc.) decreases. In that case the specific energy required and mean flow stress rises and the material shows obvious signs of the basic inhomogeneous character of strain. As a result, active shear planes are evident in a free surface and can be observed at back free surface of chip.

When the ratio of average grain size to uncut chip thickness approaches the unit size effect becomes relevant. As a result, chip formation takes place by breaking up of the individual grains of a polycrystalline material [3,7,11]. Considering that in micromachining uncut chip thickness can be even smaller than the average grain size, most polycrystalline materials are thus treated as a collection of grains with random orientation and anisotropic properties [6,7,12,4].

The crystallographic orientation affects the chip formation, shear strength and the subsurface crack generation [6,13]. The variation in shear strength causes cutting force variation over different cutting direction which results with the material induced vibration, in addition to machine induced vibration, causing degraded surface quality. To et al. [14] obtained the effects of the crystallographic orientation and the depth of cut on the surface roughness by conducting the diamond turning of single-crystal aluminium roods (Figure 6). To avoid the crystallographic effects of grains, Furukawa et al. [15] suggested the use of about ten times larger depth of cut than the average grain size.

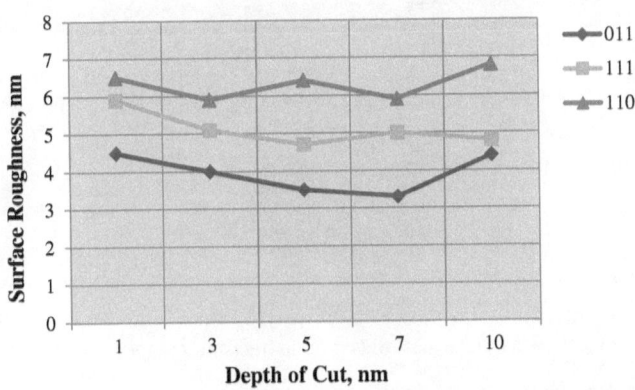

Figure 6: The effects of the crystallographic orientation and the depth of cut on the surface roughness (adapted from [7]).

Changing crystallography (multi phases or multi grains) also affects the cutting mechanism [6,3,2,4]. When the cutting tool engages from one metallurgical phase to another, the cutting conditions change, causing interrupted chip formation due to variations in the hardness of two adjacent grains. This results with variation in the cutting force and generation of additional vibration, accelerated tool wear and poor surface finish. Moreover, elastic recovery of particular grain plays important role in micromachining, especially when dealing with multiphase materials [4].

Majority of published work is dealing with work materials which are considered easy to cut, such as low hardness steels (carbon steels, high strength low alloy steels and high alloy steels which do not subject to hardening), aluminium and copper alloys, as illustrated in see Figure 7. Hardened steels, heat resistant alloys, ceramics, glasses and other hard to cut materials are less studied and seldom subject of investigation.

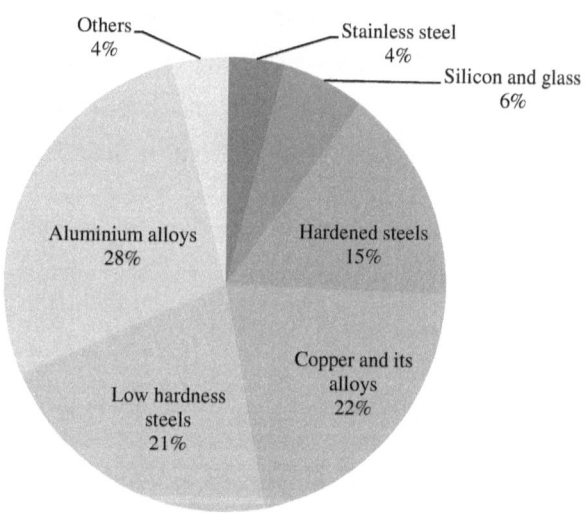

Figure 7: Typical workpiece materials used in micromachining (adapted from [3]).

2.3. Minimum chip thickness

Considering conventional machining, it is assumed that cutting tool edge is perfectly sharp and that there is no contact between the tool's clearance face and machined surface. Chip is then formed mainly by sharing of the material in front of the tool tip. However, such an assumption cannot be made for micromachining where achievable tool edge radius is commonly on the same order as the chip thickness (cutting depth). Where in conventional machining shear takes place along shear plane, in micromachining shear stress rises continuously around the cutting edge [2,7] and material seems to be pushed and deformed rather than sheared [16,17]. Therefore, micromachining processes are greatly influenced by the ratio of the depth of cut to the cutting edge radius causing a significant influence to the cutting process by a small change in the depth of cut. This ratio defines the active material removal mechanism such as cutting, plowing, or sliding and thus the resulting surface quality.

The definition of minimum chip thickness is the minimum undeformed chip thickness below which chips may not form [2,7]. Figure 8 illustrates the chip formation with respect to the cutting tool edge radius (Re) and the uncut chip thickness (h). When the uncut chip thickness is smaller than the minimum chip thickness (h_m), as shown in Figure 8(a), only elastic deformation occurs and no workpiece material will be removed by the cutter. As the uncut chip thickness approaches the minimum chip thickness (Figure 8(b)), chips are formed by shearing of the workpiece, with some elastic deformation still occurring. As a result, actual depth of cut is less than the desired depth. However, when the uncut chip thickness is larger than the minimum chip thickness (Figure 8(c), elastic deformation is significantly reduced and the entire depth of cut is removed as a chip.

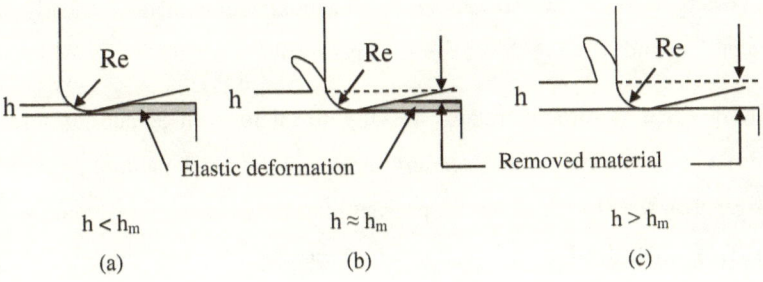

Figure 8: Schematic diagram of the effect of the minimum chip thickness (adapted from [2]).

Knowledge of the minimum chip thickness is essential in the selection of appropriate machining parameters to ensure a proper cutting and avoid plowing and sliding of the tool [6,2,7]. It is very difficult to directly measure the minimum chip thickness during the process, in spite of knowing the tool edge radius, so it is obtained by experimental results or trough numerical simulations. Minimum chip thickness depends primarily on the ratio of uncut chip thickness to cutting tool edge radius (cutting edge sharpness) and secondarily on the workpiece material properties [6] and the friction between the tool and workpiece material. Estimation of the minimum chip thickness is one of the present challenges in micromachining. Furthermore, minimum chip thickness cannot be expressed as precise and single value but rather as a range of values with unclear limits [17]. Depending on the material, minimum chip thickness was estimated to be between 5% and 40% of the tool edge radius [6,2,16,17].

2.4. Cutting forces

The majority of researchers who have investigated micromachining processes have used cutting force for monitoring or improving the quality of machined products. Excessive cutting force limits the accuracy and the depth of cut due to deflection of tool and work piece, defines the bending stress that determines the feed rate and introduce the built-up edge (B.U.E.) [2,4] [Weu01]. Therefore, reducing the cutting force in micromachining operations significantly improves material removal productivity, decrease tool deflection and tool wear, delay tool failure, and narrow workpiece tolerance limits. As in conventional machining, micromachining cutting force consists mainly of normal and tangential components, usually called shearing/cutting and plowing/thrust force, respectively.

The cutting force is directly related to chip formation. Since cutting force also determines the tool deflection and bending stress as mentioned, the tool edge radius is often larger than the chip thickness to prevent plastic deformation or breakage of the tool [2]. This small depth of cut results with large negative rake angle as shown in Figure 9. In that case workpiece is mainly processed by cutting edge causing an increase in friction on the rake face of the tool and significant elastic recovery of the workpiece along the clearance face of the tool, thus increasing the specific energy. Therefore, high ratio of the normal to the tangential component is observed as uncut chip thickness decreases as illustrated in Figure 10, which indicates a transition of material removal process from cutting to ploughing [6,7].

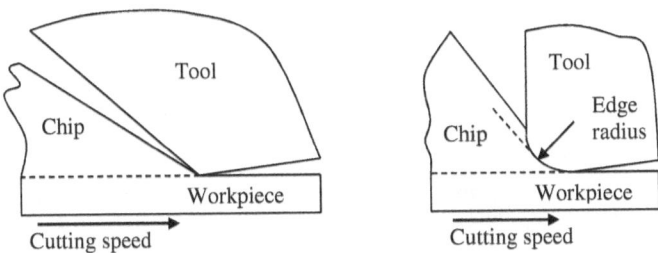

Figure 9: Schematic representation of the negative rake angle in orthogonal cutting (adapted from [17]).

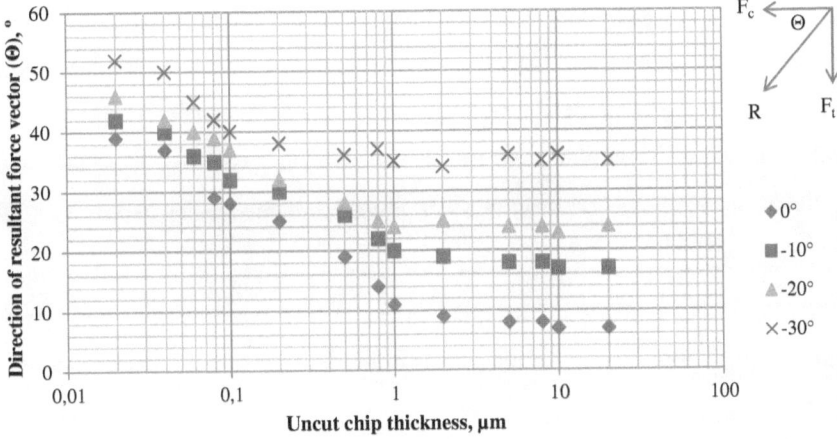

Figure 10: Resultant force vector versus uncut chip thickness at various rake angles (adapted from [7]).

Cutting force in micromachining is also significantly influenced by problems that are generally minor in macro-domain such as tool wear, unbalance (run-out) and instability (chatter) [6,3,2,7,18]. Accelerated tool wear results from increased friction between the tool and the workpiece because of the small uncut chip thickness and large negative rake angle. The smaller the uncut chip thickness, the greater the impacts on the tool wear and cutting force, i.e. cutting energy (Figure 11).

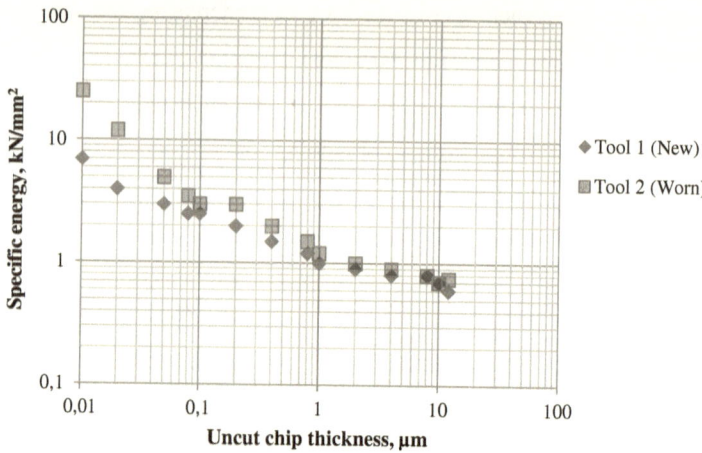

Figure 11: Specific energy versus uncut chip thickness for new and worn diamond tools (adapted from [7]).

Tool run-out is caused by tool deflection and a misalignment of the axis of symmetry between the tool and the tool holder or spindle. In macro-machining it often ignored, as the diameter of cutting tools is relatively large compared to the tool run-out and the speed is relatively slow compared to micro-machining. Tool run-out contribute to significant noise in force measurements, surface roughness and severe vibrations which causes burr formation.

Chatter introduces excessive vibrations that can lead to catastrophic failure and burr formation as a result of interaction between the dynamics of the machine tool and workpiece.

Additionally, laser assisted micromachining or vibration assisted micromachining can be applied when machining difficult to cut, hard materials, in order to reduce cutting force and extend tool life [6,3,2]. Micromachining forces and tool wear can be drastically reduced by focusing a laser beam ahead of the cutting path. This novel approach was reported by Ding et al. [19] and Kumar et al. [20]. Although the usage of laser assisted micromachining provides more consistent tool life behaviour, Kumar et al. reported larger burr heights and poorer surface finish and attribute it to effects of thermal softening. Ultrasonic vibration machining were introduced by Kumabe

[21], and later improved by Moriwaki et al. [22,23] through elliptical ultrasonic vibration (Figure 12), which showed improved cutting performance and surface quality. Vibration assisted micromachining also improves machining of ferrous metals with diamond tools by means of reduced tool wear [4]. Figure 13 shows a difference in chip formation due to no vibration, forced vibration and regenerative chatter.

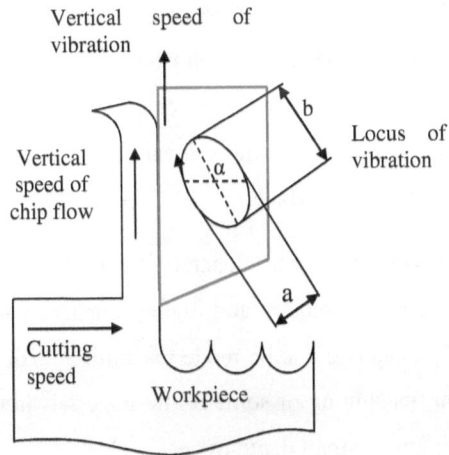

Figure 12: Principle of elliptical vibration cutting (adapted from [6]).

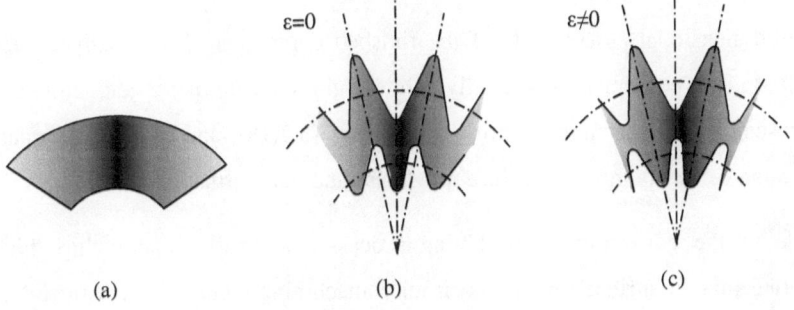

Figure 13: Chip generations due to (a) no vibration, (b) forced vibration and (c) regenerative chatter (adapted from [2]).

2.5. Brittle and ductile mode machining

Although brittle materials (such as many optical glasses, ceramics, etc.) are normally machined using conventional processing techniques such as polishing, micromachining can bring many advantages due to increased flexibility in geometries produced, greater surface finish quality, and higher material removal rate, translating to higher production throughput [6,7]. However, machining brittle material at high depth of cut has a tendency to generate excessive surface and subsurface cracking.

Shimada et al. [24] find out that, regardless of the material ductility, there exist the critical depth of cut, which causes translation from brittle to ductile material removal mechanism. Therefore, any brittle material can be machined in ductile mode if undeformed chip thickness is below the critical depth of cut, resulting with good surface finish and an uncracked surface.

The value of critical depth of cut depends on tool geometry and machining conditions. Excessive cutting velocity and higher negative rake angle increases the critical depth of cut [7], causing ductile mode machining difficult to obtain at higher feed rates. In addition, machining of some brittle materials in ductile mode is rather challenging due to extremely small depth of cut.

2.6. Surface quality

Three dimensional assessments of the finished components are usually carried out using optical equipment (especially white light interferometry and atomic force microscopy) and scanning electron microscopy [3,7,18], and the surface quality is evaluated generally through surface roughness and burr formations.

While in the conventional machining processes a smaller uncut chip thickness generate smaller surface roughness, at micromachining there exist a critical depth of cut below which surface roughness starts to increase. This phenomenon shows a strong influence of size effects on surface generation, i.e. when unit removal size decreases, issues of tool edge geometry, cutting parameters and workpiece material properties becomes dominant factors with strong influences on resulting accuracy, surface quality and integrity of the machined component. Figure 14, obtained by [16], clearly shows the effects of size effects to the surface roughness, that is influence of the ratio of feed rate to tool edge radius (a/r) on the surface roughness. Therefore, optimal depth of cut depends highly on the degree of the size effects and for that depth of cut the best surface finish is produced.

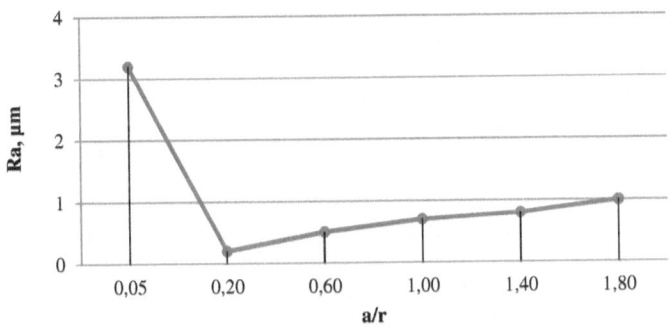

Figure 14: Experimental findings on surface roughness at the varying ratio of feed rate to tol edge radius (adapted from [16]).

Many research paper [6,3,2,7,8,4] associated optimal depth of cut with the minimum chip thickness, because below this threshold plowing and sliding effects tends to dominate machining mechanism producing discontinuous chips, bigger burr size, rough surface and elastic recovery of the workpiece material. As mentioned before, minimum chip thickness is a function of the ratio of uncut chip thickness to cutting tool edge radius and the workpiece material properties such as hardness, elastic recovery, etc., which are greatly affected by defects, impurities, grain size and crystallographic orientation, etc. Weule et al. [4] determined the achievable surface roughness of steel (SAE 1045) as a function of minimum chip thickness (and cutting edge radius). The achievable surface roughness can be predicted based on spring back of elastically deformed material as shown in Figure 15. Once the cutting depth reaches a minimum chip thickness material is removed by a shearing mechanism.

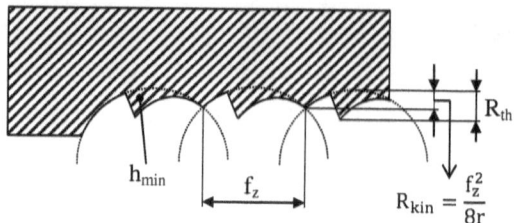

Figure 15: Theoretical surface profile based on spring back of elastically deformed material (adapted from [4]).

They also conducted experiments regarding relationship between machining parameters, material state and surface quality. Referring to Figure 16 it can be concluded that additionally to the ratio of feed rate to edge radius, cutting speed is another relevant factor which has a significant influence on surface roughness. In order to generate smaller surface roughness, higher cutting velocity and harder workpiece materials are preferable. Increased surface roughness at low cutting speeds was attributed to the formation of a built-up edge [3,4]. Mian et al. [8] in their work confirmed significant influence of cutting speed to the surface roughness and observed that the same applies for the burr root thickness.

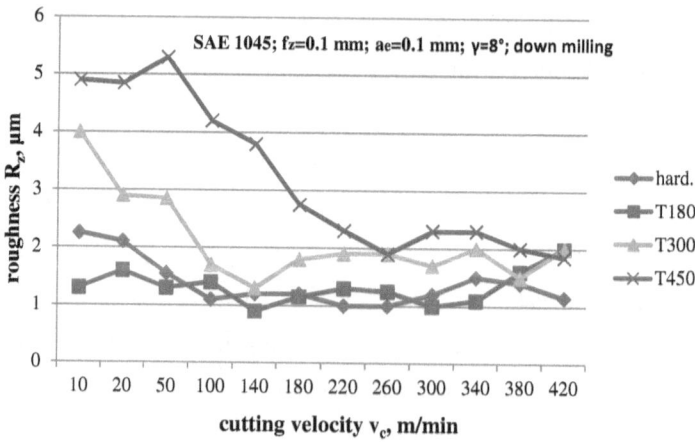

Figure 16: Influence of the cutting velocity and material state on the surface roughness (adapted from [4]).

In addition to size effects, resulting accuracy and surface quality is also directly related to the cutting tools properties and machine tools where issues such as tool wear, tool deflection, tool run-out, chatter, etc. leads to additional surface deterioration [3,2,7,8]. In order to decrease tool wear and thermal loads fluids are applied for lubrication and cooling. As fluid, either water-based emulsions or oils are used. They can be applied as a mist or flushed [18]. Flushed lubrication may be the better choice as they also improve chip evacuation process. Most unfavourable situation occurs when dealing with workpiece materials with high ductility. In that case long and continuously snarled chips are form which can easily interfere with tool engagement and burrs and contribute to poor surface quality [6,18]. Moreover, different milling strategies can also affect surface quality [3,18]. In case of machining aluminium alloy with a tungsten carbide cutter ($\Phi 800$ μm), lowest surface roughness was provided by the constant overlap spiral strategy, followed by the parallel spiral and parallel zigzag strategies [3].

Burr formation is probably the principal damage noticed on machined surfaces.

Burrs can be removed mechanically or by electro polishing. Disadvantage of mechanical approach is high manual effort or impracticability due to size of machined features [2], while electro polishing requests that no precipitations at grain boundaries or a different second phase are present [18]. Therefore, electro polishing is restricted to materials such as stainless steel, nickel and some copper base alloys. Furthermore, for monitoring purposes process must be stopped and the microstructure is evaluated by microscopy. Because there are also spots without burrs, where edges are eroded from beginning, prolonged exposure to electric field may cause rounded edges of product.

Similar to surface roughness, burr formation at micro scale is also affected with size effects. Sugawara [25] investigated the effect of the drill diameter on burr formation and concluded that burr size is reduced and cutting ability increased as drill size decreases. Generally, micromachining of ductile materials if often accompanied by burr formation, especially at the edges of microstructures [18]. When the ratio of the depth of cut to the cutting edge radius is small, high biaxial compressive stress pushes material toward the free surface and generates large top burrs [26]. Also, the kinematics of the tool as it exits from the workpiece significantly influence burr formation due to plastic deformation (i.e. bending) of chips rather than shearing [27]. Schaller et al. [28] drastically reduced burr formation by coating the surface with cyanoacrylate polymeric material. After machining, the cyanoacrylate is removed with acetone in an ultrasonic bath.

Weule et al. [4] observed that, in contrast to surface roughness, burrs most frequently occurred when cutting hard materials. It is assumed that this is a result of faster tool wear, which increase cutting edge radius leading to burr formation. Additionally, tool coatings did not result in any substantial improvement on surface roughness [3,8], while concerning burr size, best results are obtained when using tool coated with TiN, TiCN and CrTiAlN (in this order) [3].

Essentially, the relationship between surface roughness and cutting conditions is similar to that between burr size and cutting conditions. Both depend on the ratio of undeformed chip thickness to cutting edge radius, feed rate and cutting speed. However, the best process performance in terms of surface roughness and burr formation are not essentially obtained at the same cutting conditions [8].

3. Micro Cutting Tools

Materials Micro cutting tools (herein simply referred as tools) is the essential enabler for micromachining processes. Tool diameter and cutting edge radius determines achievable feature size and surface quality [7]. Cutting edge radius determines cutting tool sharpness and it influence on minimum chip thickness and determines effective rake angle of the tool as already discussed. If the diameter of micro-tools can decrease even further, the size of features on miniature components could be comparable to those produced with the lithographic techniques [2].

As far as the tool material is concerned, either tungsten carbide or single crystal diamond are used. As can be seen in Figure 17, tungsten carbide is the most common choice due to its hardness, high toughness and relatively low price [3,2,18].

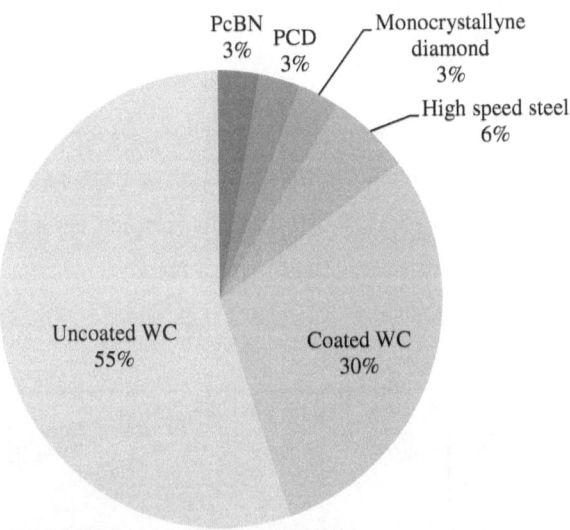

Figure 17: Principal tool materials used in micromachining (adapted from [3]).

3.1. Diamond tools

When dealing with non-ferrous and non-carbide materials, such as brass, aluminium, copper, nickel, etc., and brittle hard materials such as ceramics, silicon, glass, germanium, etc., single crystal diamond is preferred tool material due to its outstanding hardness, high thermal conductivity and elastic and shear moduli [6,2,7,18]. Furthermore, diamond tools were used in most of the early micromachining research due to their homogeneous crystalline structure which makes it easy to generate a very sharp cutting edge through grinding, e.g. a cutting edge in tens of nanometres can be achieved [7]. Lower cutting edge radius enables lower depth of cut and ensures better surface quality.

However, diamond is limited to the cutting of non-ferrous materials because of the high chemical affinity between diamond and iron. When machining ferrous materials with diamond tools, carbon of the diamond can easily diffuse, causing severe tool wear. An exception occurs in the case of low cutting speeds, when low temperatures prevent diffusion [18], or in case of vibration assisted micromachining [4].

More recently, CVD (chemical vapour deposition) diamond coated tools have become available [29]. CVD diamond tools can be used to cut tungsten carbide with a cobalt percentage of 6% or greater [7].

3.2. Tungsten carbide (WC) tools

Tools that are used to machine ferrous materials are commonly made of tungsten (wolfram) carbide (WC) [6,3,2,7,18]. Tungsten carbide cutting tools are generally used due to their hardness and strength over a broad range of temperatures (Figure 18).

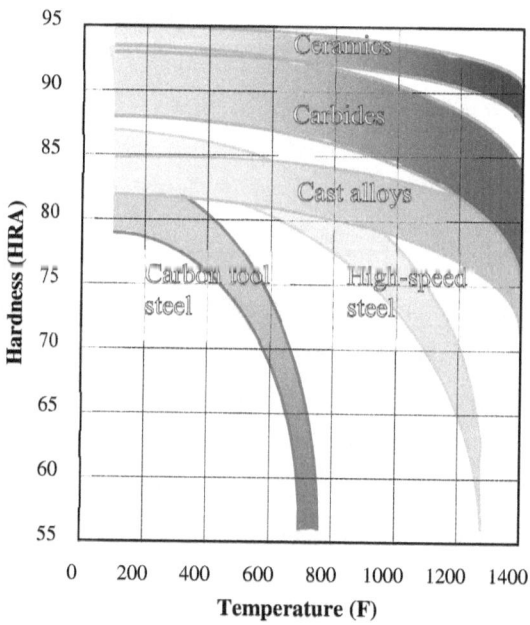

Figure 18: Hardness of cutting tool materials as a function of temperature (adapted from [2]).

In general, published literature reports tool edge radii ranging from 1 to 3 µm [3]. However, in contrast to the homogeneous crystalline structure of diamond, tungsten carbide is a hard metal composite. As a consequence, tool cutting edge is always jagged causing burr formation on ductile materials like most metals [18].

Tungsten carbide is composed of a hard phase, mainly tungsten carbide powder, and a binder phase, typically cobalt [2], but nickel and iron are also possible [18]. Tungsten carbide powder is basically responsible for tool wear resistance and it consists of submicron particles with average size of 0.2 μm [18]. Binder content and average grain size determines the mechanical properties of the tool. Low binder content results with higher tool hardness and consequently higher wear resistance, where smaller grain size is responsible for higher fracture toughness. For interrupted cut or fluctuating load, higher binder content is recommended. Furthermore, to ensure isotropic mechanical properties, cross section of the tool must consist of a sufficient number of hard particles. Therefore, according to Gietzelt et al. [18], isotropic mechanical properties of tools with diameter below 30 μm are questionable.

3.3. Coatings

Coating of tools with diameter below 0.3 mm become popular about five years ago with improvement in coating processes which enabled thinner and more uniform coating layers [18]. Main purpose of coating is to extent tool life by reducing a tool wear. In case of a thick coating, cutting edge radius is increased and consequently higher cutting forces are induced which undo the coating improvement regarding tool wear. Furthermore, formation of coating droplets must be avoided in order to prevent coating results in worse machining properties [30]. Additionally, chipping of coating layers were detected not only at the cutting edge but also in smooth substrate areas, as a result of poor adhesion of the coating.

Nowadays, TiAlN is the principal coating material applied to tungsten carbide cutters, but other coatings, such as TiN, TiCN, CrN, CrTiAlN, etc., can also be applied [3]. Majority of the coatings are quite uniform and below 1 μm in thickness, therefore rounding of the cutting edge can be neglected [18].

3.4. Tool manufacturing methods

Typically, mechanical micro grinding is used as a manufacturing process for production of micro tools. However, to achieve smaller diameters and more complex geometry, more accurate production methods may be required such as electrical discharge machining (EDM), wire electrical discharge grinding (WEDG), or focused ion beam (FIB) processes, etc. [6,3,2,7,31,32]. Figure 19 shows micro end mills developed by the focused ion beam (FIB) process, with diameter below 25 µm.

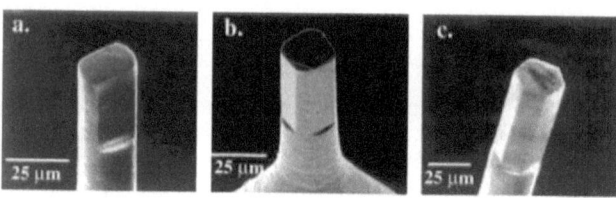

Figure 19: Micro end mills made by focused ion beam sputtering having (a) two, (b) four and (c) five cutting edges. [7]

Considering manufacturing and stability reasons, micro end mills made of single crystal diamond are no less than 50 µm in diameter [18] with achievable cutting edge radius in tens of nanometres [7]. In case of tools made of hard metals, end mills down to 20 µm in diameter [33,34,35] and drills down to 15 µm in diameter [36] are commercially available, as can be seen in Figure 20.

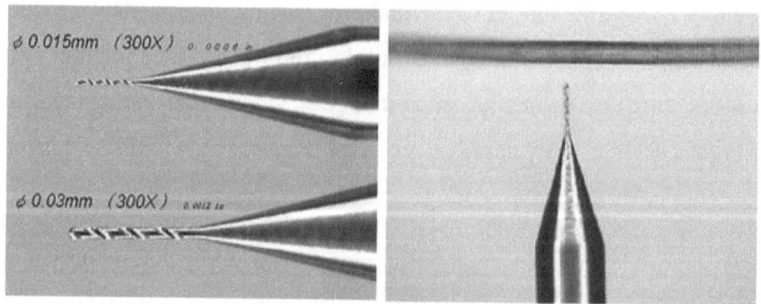

Figure 20: Left: 15 and 30 µm drill bits. Right: 20 µm drill compared with human hair. [36]

Egashira et al. [31] produced the smallest edge radius of 0.5 µm on carbide micro tool with diameter of 20 µm, using wire electrical discharge grinding process (WEDG).

Moreover, the smallest tool diameter found was 3 µm tungsten carbide tool [32]. It was also produced by wire electrical discharge grinding process [WEDG] and used for slot milling of brass workpiece, but with unpredictable performance.

3.5. Tool failure

Tool failure is another major issue in micromachining, especially when dealing with hard and difficult to cut materials such as hardened steels, heat resistant alloys, ceramics, glasses, etc. In general, the life time of micro tools is unpredictable and depends strongly on the workpiece material [3,18].

Smaller tools have decreased thermal expansion relative to their size, increased static stiffness from their compact structure, increased dynamic stability from their higher natural frequency, and the potential for decreased cost due to smaller quantities of material utilized [2,7]. However, they are also more fragile and experience larger deflection which can manifest as tool run-out and chatter marks on the workpiece. Furthermore, catastrophic tool failure may occur as a result of chip clogging, failure by fatigue or failure caused by tool wear [6,3]. Chip clogging is a result of poor chip evacuation process, and causes rapidly increase in cutting force and stress which lead to tool breakage. This mechanism is very unpredictable and happens extremely rapidly [37]. Failure by fatigue may occur as a result of tool deflection and high spindle speeds employed. Eventually, tool wear causes increase in cutting edge radius and burr formation leading to elevation of the cutting forces to levels high enough to cause failure of the tool shaft [3]. Hence, otherwise then visual inspection of the tool, tool condition could be predicted during machining based on monitoring of cutting force [6], burr formation [3] or acoustic emission [8]. Still, there is a lot of space for further work regarding this subject. Additionally, tool failure may occur as a consequence of cracks and impurities formed during manufacturing process and covered by the coating [18].

3.6. Tool design

Under micromachining, micro tools experience a different loading situation from that seen in conventional machining. To reduce tool bending and deflection, avoid the chatter marks on the workpiece and ensure stable cutting process, conventional tool design had to be reconsidered. Uhlmann et al. [38] proposed a new parametric tool design for micro end mills considering dynamic load and strain analysis trough FEM analysis. The adapted tool design has a reduced fluted length to increase the tool shaft cross section and stiffness, tapered shape with a reduced diameter at the tool peripheral edge (Figure 21) to avoid any contact with workpiece and to eliminate chatter marks on workpiece which are result of tool deflection during machining process, and rounded edge at the intersection of the constant tool shaft diameter and the conical part (Figure 22) where the bending moment is maximal (to prevent crack initiation) [6,18,38,39,40,41].

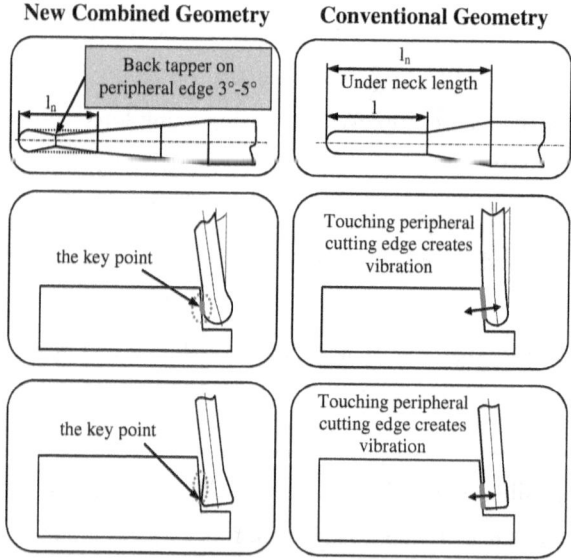

Figure 21: Cutting tool with tapered shape and reduced diameter at the tool peripheral edge (adapted from [40,41]).

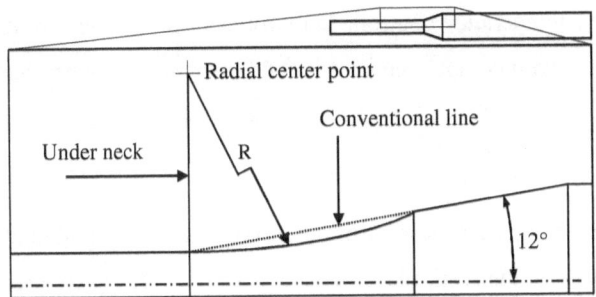

Figure 22: Cutting tool with rounded edge at the intersection of the constant tool shaft diameter and the conical part (adapted from [39]).

4. Machine tools with micromachining capability

The requirements of micro component manufacture over a range of applications are: high dimensional precision, typically better than 1 micron; accurate geometrical form, typically better than 50nm departure from flatness or roundness; and good surface finish, in the range of 10 – 50 nm [42]. To accomplish those demands, the following characteristics are required for the machine tools: high static and dynamic stiffness, high thermal stability of the frame materials, feed drives and control systems with high accuracy and short response time associated with large bandwidth and low following error for multi-axes interpolation, minimization and/or compensation of thermal effects and minimization and compensation of static and dynamic positioning errors [3]. Most of the experimental research for micromachining has been conducted on ultra-precision machine tools and machine centres or on miniaturized machine tools and micro factories built by researchers.

4.1. Ultra-precision machine tools and micromachine centres

Over the last two decades, knowledge has been accumulated for design of ultraprecision machine tools for micromachining, resulting in tough requirements such as thermal stability, precise spindle bearings and linear guides and high resolution of linear and rotary motions [6]. Currently available multi-axis controlled ultraprecision machining centres are based on conventional ultra-precision machines, operated under a temperature controlled environment [6,2]. They are used to produce small workpieces with complex geometries and microscale patterns and texture, such as moulds and dies for CD pickup lenses, contact lenses, Fresnel lenses, etc.

4.1.1. Machine materials

The stability and damping behaviour of the machine are important to avoid vibrations and chatter marks on the work piece surface as well as additional stress of the micro tool due to vibrations. Thermal and damping properties are mostly determined with materials used to produce machine components, such as the machine base, column, worktable, slide, spindle cases and carriages. A constant room temperature within 1 kelvin and absence of direct solar irradiation are advised [18].

Cast iron and granite have been widely used for fabricating machine bases and slideways [7]. Recently, as a cheaper replacement for granite, polymer concrete has become popular for ultra-precision machine tools where light weight with high damping capacity (much better than cast iron) and rigidity is required. Structural materials with a low thermal expansion coefficient and high dimensional stability have also found its application, including super-invar, synthetic granite, ceramics and Zerodur [6,7,18].

According to Gietzelt et al. [18], the shape and fixing position of the clamping to the machine also have a high impact on thermal drift due to the high thermal coefficient of expansion. For this reason, Invar and granite are most commonly used as a clamping material because of low thermal shift.

4.1.2. Spindle bearings and linear guides

To maintain acceptable productivity, micromachining requires very high speed spindle speeds due to small tool diameters and thus the dynamic characteristics of the spindle dominate machining quality. Most conventional precision machine tools are equipped with bearings and guides based on direct mechanical contact, such as ball or needle roller bearings or guides [6]. These machines are capable of producing micromachining features, but cannot achieve optical surface quality. Nowadays, aerostatic and hydrostatic bearings or guides are most commonly used [6,3,7,42,43]. Due to absent of direct mechanical contact, they introduce very little or no friction and are capable of high rotational speed with high motion accuracy. Aerostatic bearing are normally better than other bearings [42], and widely used for spindles in machine tools with medium and small loading capacity. They usually have lower stiffness than oil hydrostatic bearing spindles, but they have lower thermal deformation and their stiffness can be increased by using magnets as a preload [3,7] or with squeezed oil film dampers [42]. Hydrostatic bearing spindles are more suitable for large and heavily loaded machine tools and where very good damping properties are required.

Often to achieve higher speeds, ultra-precision machine tools are retrofitted with high-speed spindles that fit in the conventional tool holder interfaces [2] and mostly, three jaw chucks are used [18]. In that case a number of interfaces from tool to the spindle are adding up and a small deviations in the spindle may cause large run-out and result with the poor stiffness of micro-tools. Measurement of true running accuracy is a must in this case for ensuring a constant engagement of the normally two cutting edges of a micro end mill. For minimization of the run-out it is favourable to use vector controlled spindles to ensure the same orientation of the chuck inside the spindle [18]. Run-out deviation for the main spindle should be inferior to 1 µm [3].

4.1.3. High resolution of linear and rotary motions

Linear direct drive motors and piezoelectric actuators are commonly used in ultra-precision machine tools [6,3,2,7,42,43]. Compared to conventional drive mechanisms operated by friction drives, linear direct drive motors and piezoelectric actuators have no accumulative errors from friction and the motor-coupling, no loss of accuracy due to wear, and no backlash [2].

Friction drives have a long stroke and usually consist of a driving wheel, a flat or round bar and a supporting back-up roller. They offer low friction force, smooth motion, and good repeatability and reproducibility due to elastic deformation induced by preload [7].

Linear-motor direct drives (AC or DC), usually also have a long stroke and they offer better stiffness, acceleration, speed, motion smoothness, repeatability and accuracy. [44].

Piezoelectric actuators usually have a short stroke with high motion accuracy and wide response bandwidth. They have been employed in fine tool positioning so as to achieve high precision control of the cutting tool (e.g. a diamond cutting tool) [7].

A 5-axis ultraprecision micromachine centre, using aerostatic bearings and driven by linear direct drive motor, can achieve spindle rotation speed of 200000 r/min [3,42] with rotational resolution of 0.00001 degree, and the axes responsible for feed and depth of cut can achieve translational resolution of 1 nm and slideway straightness of about 10 nm/200mm [6].

4.1.4. Computer Numerical Control (CNC)

A numerical control is necessary to achieve smooth tool movements without changes in the feed rate, responsible for high accuracies of micro-structures. Following the invention of Computer Numerical Control (CNC) in the early 1970s, many companies started to develop their control systems for machine tools. The control system typically includes motors, amplifiers, switches and the controller. High speed multi-axis CNC controllers play an essential role in efficient and precision control of servo drives, error compensation (thermal and geometrical errors), optimized tool setting and direct entry of the equation of shapes [7,42].

The NC unit of the machine must be able to process sufficient numbers of instructions per second. The dynamic behaviour, namely the acceleration of the axes, the velocity to the NC-control unit and the maximum number of instructions per seconds are important to maintain a programmed feed rate. In this context, also the definition of how accurately the machine has to meet the calculated tool path is important. If the tolerance is very low, the servo-loop can cause an extreme breakdown of the feed rate. This leads to squeezing of the cutting edges, increased tool wear or even tool rupture. In the last decade, the acceleration could be improved from about 1.2 m/s to more than 20m/s (2G) by using hydrostatic drives [18].

Advanced PC-based control systems are commonly being used in the majority of commercially available ultra-precision machines as they can achieve nanometre or even sub-nanometre levels of control resolution for ultra-precision and micro-manufacturing purposes [7].

4.1.5. Position measurement and process monitoring

A major advantage of micromachining is its ability to fabricate increasingly smaller features reliably at very high tolerances. Sensor-based monitoring yields valuable information about the micromachining process that can serve the dual purpose of process control and quality monitoring, however, a high degree of confidence and reliability in characterizing the manufacturing process is required for any sensor to be utilized as a monitoring tool [6]. Figure 23 illustrates several different classes of sensors and their applicability to level of precision and type of control parameter.

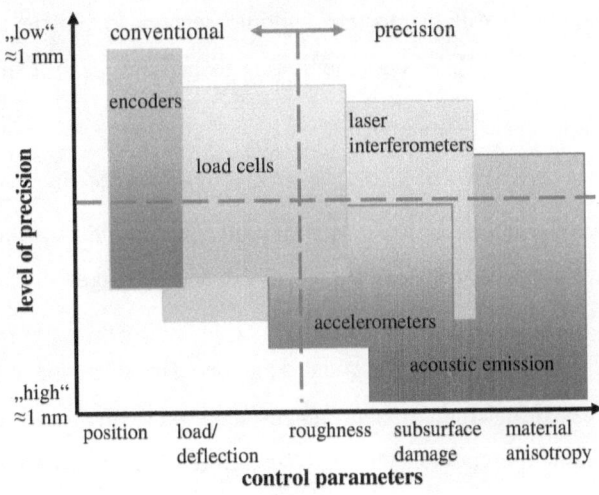

Figure 23: Sensor application vs. level of precision and control parameters (adapted from [6]).

Because of high resolution related to interferometers and ability to eliminate Abbe errors, laser encoders are suitable for ultra-precision position measurement [6,7,43]. They have a typical resolution of 20 nm, while some laser holographic-linear scales can achieve resolution of better than 10 nm [7]. Another alternative are high resolution optical encoders which can provide resolution close to that of laser encoders, but in a more industrially feasible and simple manner [7,42].

Process monitoring systems can be used to characterize, control, and improve micromachining process. Monitoring may be applied to parameters or variables such as temperature, cutting force, chatter, vibration, etc. Compared with the conventional machining processes, micromachining processes are usually difficult to monitor because of the associated very small energy emissions and cutting forces [7]. Furthermore, some control parameters, such as tool wear, tool breakage, tool engagement, material anisotropy, subsurface damage, etc., often cannot be directly measured or evaluated. Hence, process monitoring through acoustic emission, force and vibration signals draw a great deal of attention. While process monitoring through acoustic emission is the most appropriate to characterize micromachining process in the nanometre range [8], force signals can also be successfully engaged [3,2]. However, it is desirable to use multiple sensors to realize the smart and intelligent machine tool [7]. Process monitoring techniques are still subject of many research papers.

4.2. Miniaturized machine tools and micro factories

In general, micromachining is performed on precision and ultra-precision machine tools with conventional dimensions [6]. Precision and ultra-precision machine tools have several advantages including high rigidity, damping and the ability to actuate precisely based on precision sensors and actuators. However, the large scale and precisely controlled machining environment may add very high costs for the fabrication of miniature components [6,2]. Therefore, there has been strong interest by various research groups [45,46,47,48] for building miniaturized machine tools and micro factories capable to produce micro-size components and features. Micro factories are composed of different cells with different functionalities such as micro milling, micro drilling, micro press, etc. The advantage of such miniaturized machine tools and micro factories lies in increased flexibility, portability and economic benefits such as structural cost savings, shop floor space savings, energy reduction and performance benefits including reduction of thermal deformation, enhancement of static rigidity and dynamic stability as well [6]. Economic benefits also provide the ability to use more expensive construction materials that exhibit better engineering properties, while increased portability allows their deployment to any building or any location. For example, micro-factories may be suitable for the production of micro-components during military or space exploration applications, since the accessibility of large machine tools is very difficult [2]. One unique effort is to build a micro factory system where one or several minimized machine tools are small enough to be placed on the desktop.

As actuators either piezoelectric or linear direct drive (voice coil) actuators are used, in order to achieve sub-micrometre accuracies. They use high-speed air bearing spindles, as used in the majority of ultra-precision machines. However, there are challenges associated with the development of micro-machine tools. They require accurate sensors and actuators, which must be small enough to implant within the machines. The structural rigidity of micro-machine tools is less than those of precision machines. In addition, the micro machine tools can be excited by external disturbances; therefore, micro-factories require vibration isolation to achieve desired tolerances [2].

Majority of micro factory systems are still at the research stage, and only a few of them have so far found their way into industrial applications, but their application to high accuracy and fine surface quality are still constrained by low static/dynamic stiffness [2,42].

5. Numerical modelling of micromachining

The great challenge in micromachining is the microscopic level of the phenomena. It results with difficulties in experiments conducting, making of in-process observations and results measuring. Moreover, the cutting process itself is very complicated involving elastic/plastic deformation and fracture with high strain rates and temperature and for which material properties vary during the process. Most analytical modelling efforts are based on kinematics from empirical observation combined with classical cutting models at the macro level. Their applicability and accuracy are subjects of many limitations. Lately, during the last 25 years, modelling based on numerical relationships, often accompanied by computer simulation, has become the tool of choice for many researchers. It is still not a perfect solution because it also involves many of the same assumptions as in analytical modelling. However, computer based models can offer a reasonable insight into certain verifiable trends or guidance on empirical research to assist further understanding of the process.

5.1. FEM

The finite element method (FEM) is today without doubt one of the major numerical tools in computational science. This includes the results on the theory of finite elements that have been produced by mathematicians since the 1960s. Theory of finite elements is now available for very wide range of applications. But it has one critical limitation for microforming [6]. FEM is based on principles of continuum mechanics. Hence, material properties are defined as bulk material properties whereas in reality, the material behaves discontinuously in many cases of microforming. Thus, the combination of grain size and the position of a grain within the specimen have been identified as one of the main reasons for the occurring size-effects [49,50,51,52]. Due to this effect, small work pieces can no longer be regarded as homogenous continuum for simulation purposes of microforming, without some interventions. In most cases of isotropic micromachining, FEM can still be an attractive modelling method because the process can be reasonably treated in continuum space, as shown in Figure 24.

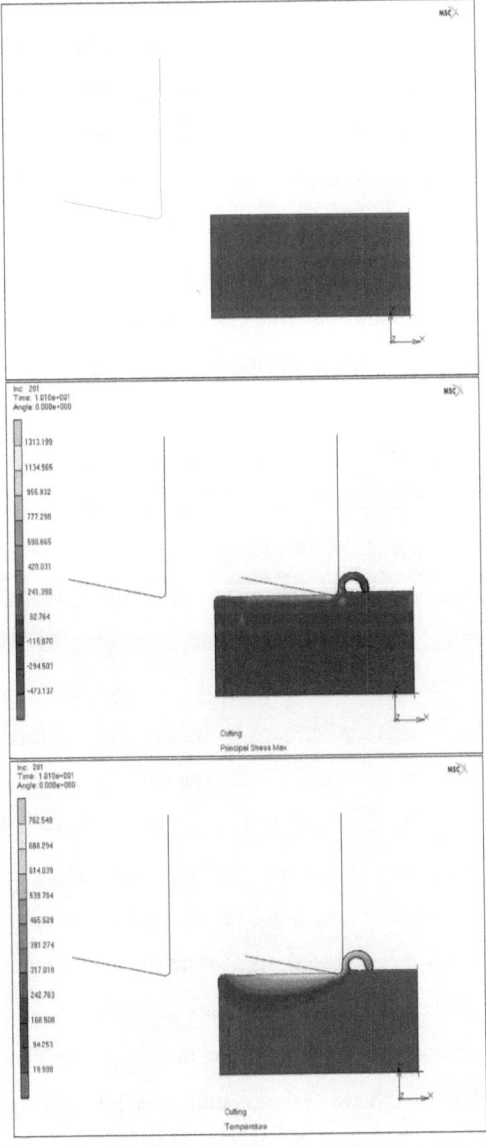

Figure 24: Finite Element Simulation of chip formation, stress and temperature distribution in machining process

In the early stages of computer simulation of metal cutting, a FEM approach was not directly used to develop a model. Rather, it was used for intermediate steps to obtain certain values using semi-mechanistic or empirical models. Ueda et al. [53] proposed such an approach in order to analyse the material removal mechanisms in micromachining of ceramics FEM was used only to calculate the J-integral around a crack in front of the cutting edge. The obtained value was then used to determine likelihood of fracture in micromachining. Using this approach, they were able to model ductile and brittle cutting modes and the results were used to maintain the process in the ductile mode. Ueda and Manabe [54] modelled chip formation in micromachining of amorphous material using rigid-plastic FEM (RPFEM). Again, they used FEM only for further understanding of localized adiabatic deformation. The model was able to produce a lamellar structure of the chip formation, which was also observed during experimental machining using a SEM. The formation of lamellar structured chip was due to the periodical occurrence of a localized shear band and smooth chip formation, the frequency of which was proportional to the depth of cut. Moriwaki et al. [55] developed a similar model using RPFEM for micro-orthogonal cutting of copper. They used a two-step FEM approach; first, the model obtains the deformed status of cutting and, second, then obtains specific values such as stress and strain which were used as input values to the thermal FE analysis combined with semi analytical formulations. The roundness of the tool edge was also taken into consideration in the model. The analysis showed that cutting ratio is decreased with an increase in the ratio of the tool edge radius to the depth of cut. It was also found that the temperature gradient in the work piece increases in front of the cutting edge due to the material flow relative to the cutting tool. Similarly, Fleischer et al. [56] used experimental values to improve a FE model to predict cutting forces in micromachining. The aforementioned FE modelling is primarily for isotropic micromachining where no crystallographic effects were considered. Chuzhoy et al. [57,58] developed a FE model for micromachining of heterogeneous material.

Wide application of FEM and commercial codes (such as Deform – 3D, Deform – 2D, Abaqus,) in scientific research begins in second half of 2000s especially after 2008 and the typical modelled process is micro milling[1]. Some of the most representative researches are listed below.

In 2008 Laia et al. [59] presented mechanisms studies of micro scale milling operation focusing on its characteristics, size effect, micro cutter edge radius and minimum chip thickness. They created a finite element model for micro scale orthogonal machining process, considering the material strengthening behaviours, micro cutter edge radius and fracture behaviour of the work material. An analytical micro scale milling force model is developed based on the FE simulations using the cutting principles and the slip-line theory.

In 2009 Yu et al. [60] research the size effect caused by the cutting edge radius and few microns per tooth in micro-milling process in order to determine the minimum thickness of cutting under different cutting condition of aluminium alloy materials 2A12 of micro-milling. Through finite element analysis, the ratio of minimum radius of thickness to the cutting edge tool radius under different conditions of cutting speed and cutter blade, the size effect, stress field and cutting force under different cutting depth were obtained.

In 2010 Afazov, Ratchev and Segal [61] presented a new approach for predicting micro-milling cutting forces using the finite element method (FEM). The trajectory of the tool and the uncut chip thickness for different micro-milling parameters (cutting tool radius, feed rate, spindle angular velocity and number of flutes) are determined and used for predicting the cutting forces in micro-milling. A number of FE analyses (FEA) are performed at different uncut chip thicknesses (0–20 μm) and velocities (104.7–4723 mm/s) for AISI 4340 steel. Based on the FE results, the relationship

---

[1] Science Direct data base reports an increasing number of scientific papers: 2007 – 1, 2008 – 5, 2009 – 4, 2010 – 5, 2011 – 10, 2012 – 16, 2013 - 45

between the cutting forces, uncut chip thickness and cutting velocity has been described by a non-linear equation proposed by the authors.

In 2011 Jin and Altintas [62] presented a slip-line field model which considers the stress variation in the material deformation region due to the tool edge radius effect. Finite element analysis software ABAQUS/Explicit is used. The ALE technique is applied to simulate the material flow during the cutting process. The total cutting forces are evaluated by integrating the forces along the entire chip-rake face contact zone and the ploughing force caused by the round edge.

In 2012 Malekian et al. [63] investigate a minimum uncut chip thickness (MUCT) under which the material is not removed but ploughed, resulting in increased machining forces that affect the surface integrity of the work piece. Analytical models based on identifying the stagnant point of the work piece material during the machining have been proposed. Based on the models, the MUCT is found to be functions of the edge radius and friction coefficient, which is dependent on the tool geometry and properties of the work piece material.

In 2013 Thepsönthi and Özel [64] studied an improvement of carbide micro-end mills performance by applying cubic boron nitride (cBN) coating. Experiments and finite element method (FEM) based simulations were used to study the effect of cBN coated tool in micro-machining of Ti-6Al-4V titanium alloy.

5.2. MD

Since the molecular dynamic (MD) simulation technique is based on interatomic force calculations, it can accommodate micro-material characteristics as well as dislocations, crack propagations, specific cutting energy, etc. Hence, many researchers have turned to MD for micromachining studies. However, there are three major obstacles in MD modelling. The core part of MD requires good representation of interatomic forces among various combinations of atoms involved in cutting, referred to as a potential. Second, since MD calculates interatomic forces among all atoms within a certain boundary, intensive computational power is required.

Therefore, many MD studies are limited to a very small space, such as at a nanometer or angstrom level. Third, MD analysis lacks a good representation of continuum behaviour of material. Therefore, most MD simulations clearly state the boundary of application [6].

The start of research work on molecular dynamic simulation of cutting can be located in early 1990s. At the beginnings researchers used copper as a work material because of its well established structure and potential function. A diamond was used as a cutting tool since it can be reasonably assumed to have a very sharp edge needed at MD level [65,66,67,68]. The work of Inamura et al. [67,69,70] focused on a trial of molecular dynamics at an atomic level cutting simulation with a couple of potential functions. This computational study showed that MD is a possible modelling tool for the microcutting process. The simulation was able to correlate the intermittent drop of potential energy accumulated in the work piece during cutting with the heat generation associated with plastic deformation of the work piece and impulsive temperature rise on the tool rake face. In a simulation of cutting of polycrystalline copper, the plastic deformation first initiated at the grain boundaries and then propagated into neighbouring grains in the direction of dislocation development. They also reported that the rate of energy dissipation in plastic deformation at this scale is larger than in conventional cutting and that a concentrated shear zone did not appear, contrary to what is normally observed in conventional cutting. Shimada et al. [71] developed a MD model for understanding the chip removal mechanism in micromachining of copper with depths of cut down to 1nm. The model was able to continuously generate dislocations in front of the tool tip and produced chip morphology showing very good agreement with experimental results. In another work [72], they also investigated crystallographic orientation effects on plastic deformation on single crystal copper using MD simulation. Because MD simulation requires impressive computational power in order to model a cutting process, many MD models have been applied to two dimensional orthogonal cutting with a very small model size, or unrealistically high cutting speed. Komanduri et al. [73]

proposed a new method called a length-restricted molecular dynamics (LRMD) simulation by fixing the length of the work material and shifting atoms along the cutting direction and applied it to nanometric cutting with realistic cutting speeds. They also studied the effect of tool geometry using several ratios of tool edge radius to the depth of cut with various parameters such as cutting force, specific energy, and subsurface damage [74]. They successfully simulated burr formation on a ductile material and crack propagation in brittle material [75]. Another typical assumption use most MD simulations of nano or micro level cutting is that the tool is rigid body and therefore dynamic change of tool geometry due to wear during cutting can be ignored. Cheng et al. [76,77] used MD simulation to model the diamond tool as a deformable body. Their model includes heat generation due to cutting and provided analysis of the relationship between the temperature and sublimation energy of the diamond atoms and silicon atoms. Fang et al. [78] investigated surface integrity of nanomachining and nanoindentation on brittle materials using a conventional MD technique. Cai et al. [79] investigated the characteristics of "dynamic hard particles" and their relationship with the diamond tool groove wear through MD simulation of nanoscale ductile mode cutting of monocrystalline silicon with diamond tools. The results show that during the cutting process, due to the work piece material phase transformation from monocrystalline to amorphous, which results in the existence of silicon atom groups with shorter bond lengths in the chip formation zone, "dynamic hard particles" having a dynamic and uneven distribution over the entire chip formation zone are formed. The distribution changes over time and cutting stages. In 2009, MD simulations were conducted to investigate the nanoscratch behaviour of nickel [80]. It has been found that the indenter shape significantly influences the nanoscratch deformation. However, the blunt indenter exhibits smaller scratch hardness and smaller ratio of the scratching force to the vertical force.

In 2011 Yang et al. [81] analysed material removal mechanism in Electrical discharge phenomena (EDM) using MD. It was found that material removal mechanism in EDM can be explained in two ways; one by vaporization and the other by the bubble explosion of superheated metals. It was also found that the metal removal efficiency is 0.02–0.05, leaving most of the melted pool resolidified. In addition, the influence of power density on the removal process was investigated, and the results showed that as the power density increases, the diameter and depth of the melted area increase, as does the metal removal efficiency.

To investigate the nanometric cutting mechanism and removal processes of materials, molecular dynamics method is utilized to conduct single crystal copper nanomachining processes simulation [82]. Nanocutting mechanism and removal processes are analysed by revised centro-symmetry parameter method. The results show that the cutting force of (111) orientation work piece are largest, and that of (110) orientation work piece is lest in the nanocutting process of single crystal copper. The average energy curves of the subsurface atoms in the work piece are trending to rapidly ascend with local decreasing in the nanocutting process, and that in the work piece show the decreasing trendy when the tool withdraw the work piece. At the same time, the peak of the average energy curves of the subsurface atoms increase with the increasing cutting depths.

In 2013, Molecular dynamics (MD) methods are applied to investigate the atomistic reaction at the tool/work piece surface to clearly expose the ductile transition response of this nanometric process [83].

Science Direct reports 107 scientific papers in MD simulation of micro cutting in 2010 and 201 paper in first seven month of 2014. These numbers are the best indicator in relevance of this research area.

5.3. FEM and MD

More recently, multi-scale modelling techniques combining FEM and MD have been proposed to overcome the disadvantages of each method and allow the coverage of a wider range of behaviour [6].

Inamura et al. [69] used multiscale modelling techniques to cover both atomic and continuum levels of simulation. Using MD, they calculated displacements of interacting atoms in cutting and then transferred into a point in the continuum by weighted mean values of the surrounding atoms to obtain continuum property values such as stress and strain. In other research [84], they found that a very complicated stress state in the work piece material including a concentrated compressive and shear strain in the primary shear zone, tensile strain along the rake face, and no shear stress inside the work piece as part of the primary shear zone, exists. They attributed this partially to buckling deformation but left a detailed explanation for future work.

6. Conclusion

The aim of this review article is to summarize existing knowledge and highlight current challenges, restrictions and advantages in the field of micro manufacturing. Although natural curiosity and industry demands are responsible for active research in this field for some time, particular issues and challenges still exists. Additional research motivation lies in bridging the knowledge gap between materials at the macro and micro scale.

The macro and micro machining processes share the same material removal principle and there are many similar issues between them, such as regenerative chatter, tool wear, monitoring strategies, etc. However, owning to the inevitable size effects, the direct knowledge transfer to the micro domain by pure scaling is not possible and many assumptions which are taken for granted in macro domain are not valid in micro domain. Hence, further research is required in order to fully understand micromachining process mechanics which is primarily influenced by grain size and different grain properties in case of multiphase materials and requires extensive research in chip removal processes and material properties.

Substantial advance in micromachining field can be evident from development of cutting tools and machining tools. Tungsten carbide material with micro grain size allows production of cutting tools with smaller cutting edge radius which enables the lower values of uncut chip thickness. Furthermore, redesigned tool geometry offer higher tool stiffness, and improved tool coatings (uniform and thin) provide tool wear reduction and longer tool life. However, micromachining of brittle and very ductile materials is still a challenge regarding reasonable surface quality. While ductile materials introduce bigger burr size, brittle materials cause low material removal rate and high tool wear. Burr formation is the most critical aspect regarding quality of the machined product and it is influenced by the material properties and the machining parameters and strategies. In order to assure more consistent tool life cutting forces encountered during micromachining can be reduced by employing novel approach such as laser or vibration assisted micromachining.

Although conventional machine centres are capable for micromachining processes, full advantage of micromachining benefits can be accomplished by employing machine tools specially designed for this purpose. Furthermore, from the last decade there exists a strong interest in building miniaturized machine tools and micro factories with micromachining capability. The advantages of such miniaturized machine tools and micro factories are flexibility, mobility and various economic. Regarding process monitoring techniques, acoustic emission stands out among the force and vibration signals monitoring. Current researches regarding acoustic emission are oriented at improving the prediction of tool failure, surface finish and burr formation.

Although mechanical micromachining processes still demands various improvement, mostly regarding higher material removal rates and selection of process parameters in order to achieve stable cutting process, compared with other microfabrication techniques (i.e. MEMS) its benefits lies in low cost production, small batch sizes and capability to produce accurate 3D free-form surfaces in a variety of metallic alloys, composites, polymers and ceramic materials.

This investigation of the existing knowledge in the field of micromachining surely leads to question of the possible developing directions. The trace need to be searched in mentioned challenges that are still waiting to be coped with. Nevertheless those challenges demand a foregoing development of necessary infrastructure in the form of advanced gauging which can result in better following of process parameters and in some new knowledge of their improvement.

The literature is impressive, especially that related to modelling and process physics. Hundreds of scientific researches show how great challenges it presents to scientists all over the world. There are two elemental areas that push scientists to continue their research and will probably result with a new scientific work:

- Understanding of process mechanics in the modelling process is extremely important because of "noncontinuous" nature of the materials at this scale and difficulty of validating model predictions.
- Metrology also presents a great challenge and is related to the problem of result validation.

7. Bibliography

[1] K. Cheng and D. Hou, Eds., *Micro-Cutting: Fundamentals and Applications*, 1st ed.: John Wiley & Sons Inc, 2013.

[2] J. Chae, S. S. Park, and T. Freiheit, "Investigation of Micro-Cutting Operations," *International Journal of Machine Tools & Manufacture*, vol. 46, no. 3-4, pp. 313-332, 2006.

[3] M. A. Câmara, J. C. Campos Rubio, A. M. Abrão, and J. P. Davim, "State of the Art on Micromilling Of Materials, a Review," *Journal of Materials Science & Technology*, vol. 28, no. 8, pp. 673-685, 2012.

[4] H. Weule, V. Hüntrup, and H. Tritschler, "Micro-Cutting of Steel to Met New Requirements in Miniaturization," *CIRP Annals - Manufacturing Technology*, vol. 50, no. 1, pp. 61-64, 2001.

[5] Milton C. Shaw, "The Size Effect in Metal Cutting," *Sadhana*, vol. 28, no. 5, pp. 875-896, 2003.

[6] D. Dornfeld, S. Min, and Y. Takeuchi, "Recent Advances in Mechanical Micromachining," *CIRP Annals - Manufacturing Technology*, vol. 55, no. 2, pp. 745-768, 2006.

[7] Xizhi Sun and Kai Cheng, "Chapter 2 - Micro-/Nano-Machining through Mechanical Cutting," in *Micro-Manufacturing Engineering and Technology.*, 2010.

[8] A. J. Mian, N. Driver, and P. T. Mativenga, "Identification of Factors that Dominate Size Effect in Micro-Machining," *Internationa Journal of Machine Tools & Manufacture*, vol. 51, no. 5, pp. 383-394, 2011.

[9] W.R. Backer, E.R. Marshall, and M.C. Shaw, "Size effect in metal cutting," *Transactions of the ASME*, vol. 74, no. 1, pp. 61-72, 1952.

[10] N. Taniguchi, "Current Status in and Future Trends of Ultraprecision Machining and Ultrafine Material Processing," *CIRP Annals - Manufacturing Technology*, vol. 32, no. 2, pp. 573-582, 1983.

[11] A Aramcharoen and P.T. Mativenga, "Size Effect and Tool Geometry in Micro Milling of Tool," *Precision Engineering*, vol. 33, no. 4, 2009.

[12] K. Iwata, T. Moriwaki, and K. Okuda, "Ultra-High Precision Diamond Cutting of Copper," *Memoirs of the Faculty of Engineering, Kobe University*, vol. 31, pp. 93-102, 1984.

[13] T. Sumomogi, M. Nakamura, T. Endo, T. Goto, and S. Kaji, "Evaluation of Surface and Subsurface Cracks in Nanoscale-Machined Single-Crystal Silicon by Scanning Force Microscope and Scanning Laser Microscope," *Materials Characterization*, vol. 48, no. 2, pp. 141-145, 2002.

[14] S. To, W.B. Lee, and C.Y. Chan, "Ultraprecision diamond turning of aluminium single crystals," *Journal of Mechanical Science and Technology*, vol. 63, no. 1-3, pp. 157-162, 1997.

[15] Y. Furukawa and N. Moronuki, "Effect of Material Properties on Ultra Precise Cutting Processes," *CIRP Annals - Manufacturing Technology*, vol. 37, no. 1, pp. 113-116, 1998.

[16] K. S. Woon and M. Rahman, "Extrusion-like Chip Formation Mechanism and its Role in Suppressing Void Nucleation," *CIRP Annals - Manufacturing Technology*, vol. 59, no. 1, pp. 129-132, 2010.

[17] F. Ducobu, E. Rivière-Lorphèvre, and E. Filippi, "Chip Formation in Micro-cutting," in *9th National Congress on Theoretical and Applied Mechanics*, Brussels, 2012.

[18] T. Gietzelt and L. Eichhorn, "Mechanical Machining by Drilling, Milling and Slotting," in *Micromachining Techniques for Fabrication of Micro and Nano Structures*.: InTech, 2012, pp. 159-182.

[19] H. Ding, N. Shen, and Y.C. Shin, "Thermal and mechanical modeling analysis of laser-assisted micro-milling of difficult-to-machine alloys," *Journal of Materials Processing Technology*, vol. 212, pp. 601-613, 2012.

[20] M. Kumar and S.N. Melkote, "Process Capability Study of Laser-Assisted

Micro Milling of a Hardened Tool Steel," *Journal of Manufacturing Processes*, vol. 14, no. 1, pp. 41-51, 2012.

[21] J. Kumabe, "Vibration Cutting," Jikkyou Publishing Co., Tokyo, Japan, 1979.

[22] T. Moriwaki and E. Shamoto, "Ultraprecision Diamond Turning of Stainless Steel by Applying Ultrasonic Vibration," *CIRP Annals*, vol. 40, no. 1, pp. 559-562, 1991.

[23] T. Moriwaki and E. Shamoto, "Ultrasonic Elliptical Vibration Cutting," *CIRP Annals*, vol. 44, no. 1, pp. 31-34, 1995.

[24] S. Shimada et al., "Brittle-ductile transition phenomena in microindentation and micromachining," *CIRP Annals*, vol. 44, no. 1, pp. 523-526, 1995.

[25] Akira Sugawara, "Study on Micro Diameter Drill Working: Effects of Working Conditions on Burr and Cutting Force," *Science reports of the Research Institutes, Tohoku University. Ser. A, Physics, chemistry and metallurgy*, vol. 29, pp. 122-140, 1980.

[26] G. Bissacco, H.N. Hansen, and L. De Chiffre, "Micromilling of Hardened Tool Steel for Mould Making Applications," *Journal of Materials Processing Technology*, vol. 167, no. 2, pp. 201-207, 2005.

[27] G. Byrne, D. Dornfeld, and B. Denkena, "Advancing cutting technology," *CIRP Annals*, vol. 52, no. 2, pp. 483-507, 2003.

[28] T. Schaller, L. Bohn, J. Mayer, and K. Schubert, "Microstructure grooves with a width of less than 50 mm cut with ground hard metal micro end mills," *Precision Engineering*, vol. 23, no. 4, pp. 229-235, 1999.

[29] SGS Tool Company. [Online]. http://www.sgstool.com/languages/french/catalogs/PDFsections/DIA-Carb.pdf

[30] K. Klocke, J.V. Bodenhausen, and K. Arntz, "Prozesssicherheit bei der Mikrofräsbearbeitung," *wt Werkstattstechnik online*, vol. 11-12, no. 95, pp. 882-886, 2005.

[31] K. Egashira and K. Mizutani, "Milling Using Ultra-Small Diameter Ball End

Mills Fabricated by Electrical Discharge Machining," *Journal of the Japan Society of Precision Enginering*, vol. 69, no. 1, pp. 1149-1153, 2003.

[32] K. Egashira, S. Hosono, S. Takemoto, and Y. Masao, "Fabrication and cutting performance of cemented tungsten carbide micro-cutting tools," *Precision Engineering*, vol. 35, no. 4, pp. 547-553, 2011.

[33] Paul Horn GmbH. [Online]. http://www.directindustry.com/prod/paul-horn/solid-carbide-end-mills-16174-385022.html

[34] TENAQUIP Limited. [Online]. http://www.tenaquip.com/shop/itemDetail.do;jsessionid=42FC2A29067D61B1B1C9A95DD0955849?itm_id=166569&itm_index=425

[35] SCN Industrial. [Online]. https://www.scnindustrial.com/index/itemDetail.do;jsessionid=3A7DE8C1FD7A3F7E9C34F09CA9E4D962?itm_id=166569

[36] Atom Precision of America. [Online]. http://www.atomprecision.com/aboutus.html

[37] I.N. Tansel et al., "Tool Wear Estimation in Micro-Machining. Part I: Tool Usage Cutting Force Relationship," *International Journal of Machine Tools and Manufacture*, vol. 40, no. 4, pp. 599-608, 2000.

[38] E. Uhlmann and K. Schauer, "Dynamic Load and Strain Analysis for the Optimization of Micro End Mills," *CIRP Annals*, vol. 54, no. 1, pp. 75-78, 2005.

[39] Hitachi Metals America. [Online]. http://www.hitachimetals.com/product/cuttingtools/pdf_solid/epdb_epdbp_low_res.pdf

[40] Hitachi Metals America. [Online]. http://www.hitachimetals.com/product/cuttingtools/pdf_solid/epdb_epdbp_low_res.pdf

[41] Hitachi Metals America. [Online].

http://www.hitachimetals.com/product/cuttingtools/pdf_solid/epdr-TH.pdf

[42] K. Cheng and D. Huo, "Design of a 5-Axis Ultraprecision Micro Milling Machine – UltraMill: Part 1: Holistic Design Approach, Design Considerations, and Specifications," *International Journal of Advanced Manufacturing Technology*, vol. 47, no. 9-12, pp. 867-877, 2010.

[43] O. Riemer, "Advances in Ultra Precision Manufacturing," in *Japan Society for Precision Engineering Spring Meeting*, Japan, 2011.

[44] D. Huo and K. Cheng, "A dynamics-driven approach to the design of precision machine tools for micro-manufacturing and its implementation perspectives," *Proceedings of the Institution of Mechanical Engineers, Part B: Journal of Engeering Manufacture*, vol. 222, no. 1, pp. 1-13, 2008.

[45] M. Tanaka, "Development of desktop machining microfactory," *Riken Review*, vol. 34, pp. 46-49, 2001.

[46] E. Kussul et al., "Development of micromachine tool prototypes for microfactories," *Journal of Micromechanics and Microengineering*, vol. 12, no. 6, pp. 795-812, 2002.

[47] Y. Okazaki, N. Mishima, and K. Ashida, "Microfactory - concept, history, and developments," *Journal of Manufacturing Science and Engineering*, vol. 126, no. 4, pp. 837-844, 2004.

[48] S.Y. Liang, "Performance Evaluation of a Micro-Scale Vertical Milling Machine for Precision Manufacturing," in *55th CIRP General Assembly*, Antalya, Turkey, 2005.

[49] M. Geiger, M. Kleiner, R. Eckstein, N. Tiesler, and U. Engel, "Microforming, keynote paper," *Annals of the CIRP*, vol. 50, no. 2, pp. 445-462, 2001.

[50] U. Engel, A. Meßner, and M. Geiger, "Advanced concept for the FE—simulation of metal forming processes for the production of microparts," in *Advanced Technology of Plasticity 1996, Proceedings of the Fifth ICTP*, vol. 2, Ohio, USA, 1996, pp. 903-907.

[51] U. Engel and R. Eckstein, "Microforming—from basic research to its realization," *Journal of Materials Processing Technology*, no. 125-126, pp. 35-44, 2002.

[52] Z. Keran, M. Math, and B. Grizelj, "Experimental and Numerical Analysis of Coining Process using Microforming Approach," *Tehnicki vjesnik - Technical Gazette*, vol. 18, no. 4, pp. 505-510, 2011.

[53] K. Ueda, T. Sugita, and H. Hiraga, "Integral Approach to Material Removal Mechanisms in Microcutting of Ceramics," *CIRP Annals*, vol. 40, no. 1, pp. 61-64, 1991.

[54] K. Ueda and K. Manabe, "Chip Formation Mechanism in Microcutting of an Amorphous Metal," *CIRP Annals*, vol. 41, no. 1, pp. 129-132, 1992.

[55] T. Moriwaki, N. Sugimura, and S. Luan, "Combined Stress, Material Flow and Heat Analysis of Orthogonal Micromachining of Copper," *CIRP Annals*, vol. 42, no. 1, pp. 75-78, 1993.

[56] J. Fleischer et al., "An Integrated Approach to the Modeling of Size-Effects in Machining with Geometrically Defined Cutting Edges," in *Proceedings of the CIRP Modeling Workshop*, Chemnitz, Germany, 2005, pp. 123-129.

[57] L. Chuzhoy, R.E. DeVor, and S.G. Kapoor, "Machining Simulation of Ductile Iron and Its Constituents, Part 2: Numerical Simulation and Experimental Validation of Machining," *Journal of Manufacturing Science and Engineering, Transactions of the ASME*, vol. 125, no. 2, pp. 192-201, 2003.

[58] L. Chuzhoy, R.E. DeVor, S.G. Kapoor, A.J. Beaudoin, and D.J. Bammann, "Machining Simulation of Ductile Iron and Its Constituents, Part 1: Estimation of Material Model Parameters and Their Validation," *Journal of Manufacturing Science and Engineering, Transactions of the ASME*, vol. 125, no. 2, pp. 181-191, 2003.

[59] X. Laia, H. Lia, C. Lia, Z. Lina, and J. Nib, "Modelling and analysis of micro scale milling considering size effect, micro cutter edge radius and minimum

chip thickness," *International Journal of Machine Tools & Manufacture*, vol. 48, pp. 1-14, 2008.

[60] X. M. Yu, Y. Z. Sun, and H. T. Liu, "Finite Element Simulation and Analysis of Size Effect in Micro-Milling Process," *Applied Mechanics and Materials*, vol. 16-19, pp. 1159-1163, 2009.

[61] S.M. Afazov, S.M. Ratchev, and J. Segal, "Modelling and simulation of micro-milling cutting forces," *Journal of Materials Processing Technology*, vol. 210, no. 15, pp. 2154-2162, 2010.

[62] X. Jin and Y. Altintas, "Slip-line field model of micro-cutting process with round tool edge effect," *Journal of Materials Processing Technology*, vol. 211, pp. 339-355, 2011.

[63] M. Malekian, M.G. Mostofa, S.S. Park, and M.B.G. Jun, "Modeling of minimum uncut chip thickness in micro machining of aluminum," *Journal of Materials Processing Technology*, vol. 212, pp. 553-559, 2012.

[64] T. Thepsonthi and T. Ozel, "Experimental and finite element simulation based investigations on micro-milling Ti-6Al-4V titanium alloy: Effects of cBN coating on tool wear," *Journal of Materials Processing Technology*, vol. 213, pp. 532-542, 2013.

[65] I.F. Stowers et al., "Molecular- Dynamics Simulation of the Chip Forming Process in Single Crystal Copper and Comparison with Experimental Data," in *Proceedings of the ASPE Annual Meeting*, Santa Fe, USA, 1991, pp. 100-109.

[66] N. Ikawa, S. Shimada, H. Tanaka, and G. Ohmori, "Atomistic Analysis of Nanometric Chip Removal as Affected by Tool-Work Interaction in Diamond Turning," *CIRP Annals*, vol. 40, no. 1, pp. 551-554, 1991.

[67] T. Inamura, H. Suzuki, and N. Takazawa, "Cutting Experiments in a Computer Using Atomic Model of a Copper Crystal and a Diamond Tool," *Journal of the JSPE*, vol. 56, pp. 1480-1486, 1990.

[68] S. Shimada, N. Ikawa, G. Ohmori, H. Tanaka, and U. Uchikoshi, "Molecular

Dynamics Analysis as Compared with Experimental Results of Micromachining," *CIRP Annals*, vol. 41, no. 1, pp. 117-120, 1992.

[69] T. Inamura, N. Takezawa, and Y. Kumaki, "Mechanics and Energy Dissipation in Nanoscale Cutting," *CIRP Annals*, vol. 42, no. 1, pp. 79-82, 1993.

[70] T. Inamura, N. Takezawa, and N. Taniguchi, "Atomic-Scale Cutting in a Computer Using Crystal Models of Copper and Diamond," *CIRP Annals*, vol. 41, no. 1, pp. 121-124, 1992.

[71] S. Shimada, N. Ikawa, H. Tanaka, G. Ohmori, and J. Uchikoshi, "Molecular Dynamics Analysis of Cutting Force and Chip Formation Process in Microcutting," *Journal of the Japan Society for Precision Engineering*, vol. 59, no. 12, pp. 2015-2021, 1993.

[72] S. Shimada, N. Ikawa, H. Tanaka, and J. Uchikoshi, "Structure of Micromachined Surface Simulated by Molecular Dynamics Analysis," *CIRP Annals*, vol. 43, no. 1, pp. 51-54, 1994.

[73] R. Komanduri, N. Chandrasekaran, A.N. Khajavi, and L.M. Raff, "New Method for Molecular Dynamics Simulation of Nanometric Cutting," *Philosophical Magazine B; Physics of Condensed Matter; Statistical Mechanics, Electronic, Optical and Magnetic Properties*, vol. 77, no. 1, pp. 7-26, 1998.

[74] R. Komanduri, N. Chandrasekaran, and L.M. Raff, "Effect of Tool Geometry in Nanometric Cutting: A Molecular Dynamics Simulation Approach," *Wear*, vol. 219, no. 1, pp. 84-97, 1998.

[75] R. Komanduri, N. Chandrasekaran, and L.M. Raff, "MD Simulation of Exit Failure in Nanometric Cutting," *Materials Science and Engineering A*, vol. 311, no. 1-2, pp. 1-12, 2001.

[76] K. Cheng, X. Luo, R. Ward, and R. Holt, "Modeling and Simulation of the Tool Wear in Nanometric Cutting," *Wear*, vol. 255, no. 7-12, pp. 1427-1432, 2009.

[77] X. Luo, K. Cheng, X. Guo, and R. Holt, "An Investigation on the Mechanics of Nanometric Cutting and the Development of Its Test-Bed," *International Journal of Production Research*, vol. 45, no. 15, pp. 1681-1686, 2003.

[78] F.Z. Fang, H. Wu, and Y.C. Liu, "Modelling and Experimental Investigation on Nanometric Cutting of Monocrystalline Silicon," *International Journal of Machine Tools and Manufacture*, vol. 45, no. 15, pp. 1681-1686, 2005.

[79] M.B. Cai, X.P. Li, and M. Rahman, "Characteristics of "dynamic hard particles" in nanoscale ductile mode cutting of monocrystalline silicon with diamond tools in relation to tool groove wear," *Wear*, vol. 263, no. 7-12, pp. 1459-1466, 2007.

[80] Y. Gao et al., "Molecular dynamics simulation of effect of indenter shape on nanoscratch of Ni," *Wear*, vol. 267, no. 11, pp. 1998-2002, 2009.

[81] X. Yang, J. Guo, X. Chen, and M. Kunieda, "Molecular dynamics simulation of the material removal mechanism in micro-EDM," *Precision Engineering*, vol. 35, no. 1, pp. 51-57, 2011.

[82] J. Chen et al., "Nano-Cutting Molecular Dynamics Simulation of a Copper Single Crystal," *Procedia Engineering*, vol. 29, pp. 3478-3482, 2012.

[83] O.A. Olufayo and K. Abou-El-Hossein, "Molecular Dynamics Modeling of Nanoscale Machining of Silicon," *Procedia CIRP*, vol. 8, pp. 504-509, 2013.

[84] T. Inamura, N. Takezawa, Y. Kumaki, and T. Sata, "On a Possible Mechanism of Shear Deformation in Nanoscale Cutting," *CIRP Annals*, vol. 43, no. 1, pp. 47-50, 1994.

[85] Hitachi Tool Engineering. [Online]. http://pdf.directindustry.com/pdf/hitachi-tool/epoch-micro-drill-emd/33369-395767.html

[86] R.T. Howe and R. S. Muller, "Polycrystalline Silicon Micromechanical Beams," *Journal of The Electrochemical Society*, vol. 130, no. 6, pp. 1420-1423, 1983.

[87] L.S. Fan, Y.C. Tai, and R.S. Muller, "Integrated movable micromechanical

structures for sensors and actuators," *Electron Devices, IEEE Transactions*, vol. 35, no. 6, pp. 724-730, 1988.

[88] K.S.J. Pister, M.W. Judy, S.R. Burgett, and R.S. Fearing, "Microfabricated hinges," *Sensors and Actuators A: Physical*, vol. 33, no. 3, pp. 249-256, 1992.

[89] H.C. Nathanson, W.E. Newell, R.A. Wickstrom, and J.R. Davis Jr., "The resonant gate transistor," *Electron Devices, IEEE Transactions*, vol. 14, no. 3, pp. 117-133, 1967.

I want morebooks!

Buy your books fast and straightforward online - at one of the world's fastest growing online book stores! Environmentally sound due to Print-on-Demand technologies.

Buy your books online at

www.get-morebooks.com

Kaufen Sie Ihre Bücher schnell und unkompliziert online – auf einer der am schnellsten wachsenden Buchhandelsplattformen weltweit! Dank Print-On-Demand umwelt- und ressourcenschonend produziert.

Bücher schneller online kaufen

www.morebooks.de

OmniScriptum Marketing DEU GmbH
Heinrich-Böcking-Str. 6-8
D - 66121 Saarbrücken

Telefax: +49 681 93 81 567-9

info@omniscriptum.de
www.omniscriptum.de

www.ingramcontent.com/pod-product-compliance
Lightning Source LLC
Chambersburg PA
CBHW031547210526
45464CB00003B/1193